Silvanus P Thompson
Teacher

James Greig
MSc PhD CEng FIEE FRSE

Her Majesty's
Stationery Office
London

Silvanus Thompson as President of the Institution of Electrical Engineers 1899–1900. Painted by J. Walker West, RWS.

Reproduced by kind permission of the Institution of Electrical Engineers

Silvanus P. Thompson: Teacher
born 1851, died 1916

Upbringing: First Steps in Science

The Great Exhibition of 1851 heralded a new era in British domestic history, a period distinguished by scientific achievement, technical and economic advance and social reform. To be born into a cultured Quaker family in that year would have augured well for an interesting childhood, for parents sensitive and responsive to social movements, scientific advances and industrial developments, could not fail to be stimulated and excited by the events of the next twenty years.

To the wife of Silvanus Thompson, a young master at Bootham School in York, was born, on 19th June 1851, a son who was named Silvanus Phillips. Before the boy would reach manhood, the Second Reform Bill would have become law, the *Great Eastern* would have laid a telegraph cable across the Atlantic and Thomas Henry Huxley would have made his famous retort to Bishop Wilberforce, when speaking in support of Darwin at the British Association.

The father and grandfather of the schoolmaster, both, by profession, pharmaceutical chemists, were not only enterprising in business but were men of liberal interests with wide ranging social and scientific contacts. With such a background, teaching, although not in the direct family tradition, was a not unnatural choice for the young man.

As a Quaker, the elder Silvanus Thompson was debarred from studying at Oxford or Cambridge, and he went for a time to University College London, founded in the mid-1820s with the provision that there be no 'religious tests or doctrinal forms which would oppose a barrier to the education of any sect among His Majesty's subjects'.

In 1841 he was appointed a master at the Friends' School for boys at York, known as Bootham School. Founded in 1823 as a private enterprise, this had been acquired in 1829 by a group of Yorkshire Quakers to provide for the sons of Friends an education consonant with the principles and practices of the Quaker faith.

Field work in natural history was encouraged as an extra-curricular activity in the school. It was through his interest in botany that Thompson met Bridget, daughter of John Tatham of Settle. John Tatham, a local businessman, had developed his hobby of collecting plants so far that he had become widely recognized as a botanist of some standing. In 1848 Thompson married Bridget Tatham, and they settled in a house adjoining the playing fields of Bootham School. There they reared a family of five sons and three daughters. Silvanus Phillips was their second child. The parents were enlightened educationalists in the best sense of that term and, in the years before school age had been reached, Mrs Thompson introduced the children to drawing, nature study and handicraft. A good-tempered, happy child, Silvanus responded to his mother's encouragement, and showed remarkable aptitude for drawing and for the making of models.

In 1858 the two eldest boys entered Bootham School. Silvanus was studious, though not to the exclusion of sport and the other less scholarly activities of the normal schoolboy. He appears to have progressed uneventfully through both junior and senior school. His ability for sketching

1 *Cartoon drawn by Thompson when he was a boy*

was fostered by the able and sympathetic visiting art master, Edwin Moore, and from time to time Silvanus delighted his schoolfellows with his mischievously clever caricatures. The impish enjoyment with which he indulged this propensity remained with him almost throughout his life.

When he entered the senior class he took up the study of astronomy with great enthusiasm. In the school observatory he was able to make quite extensive observations, some of which were reported to the British Association in 1867 in the *Report of the Luminous Meteor Committee*.

In June 1867 Silvanus cleared his first major educational hurdle – the Matriculation Examination of the University of London. He and several contemporaries were brought to London by one of their masters. They stayed in Angus's Hotel in a street which ran between the Strand and the river. The Victoria Embankment was under construction at the time. The examinations were held in Burlington House where the offices and examination halls of the University of London were then situated. The papers were, apparently, not in any way troublesome. In a letter to his father he wrote 'both the Latin papers have been very easy . . . Both Jim and I finished the Grammar in an hour – only one half of the specified time'. It was the boy's first visit to London. There was time to see many of the sights and these did not fail to thrill the eager young lad. The visits included one to the Polytechnic to see the famous optical illusion, *Pepper's Ghost*.

Silvanus was to follow his father in the teaching profession. He threw himself enthusiastically into the work of the teachers' training course in the Quaker Training College at Pontefract, called the Flounders Institute. The University of London was at that time purely an examining body, and the students at the Flounders Institute could, in parallel with their teacher training, prepare for the examinations of the University. Devoting himself chiefly to classical studies, Silvanus intended to sit the examinations for the London BA in the summer of 1869. In March an epidemic of typhoid broke out in the college and Silvanus was one of the victims. He had the good fortune to be at home when the attack developed and thus had the undivided attention and care of his mother throughout his illness. For some time his condition was critical, but by the autumn he had recovered enough to return to college and in due course to sit and to pass the examinations for the BA. He was nineteen, and soon after graduating, he was appointed a junior master at Bootham School, where his father had become senior master.

Almost immediately Silvanus took up the study of science and after three years he became a science master, teaching chemistry, physics and electricity. To improve his qualifications for this post he prepared to take the examinations for the London BSc, which comprised Preliminary, Intermediate and Final parts.

As a schoolmaster, he was not as popular as other masters because of his lack of interest in sport, but, for boys who enjoyed scientific experimentation or making scientific collections, or who had a taste for art, he was the unfailing source of encouragement and information, from which so often arises a deep and understanding friendship between master and pupil.

It was at this time that Silvanus began to develop his appreciation of music. There was no piano in the Thompson household. There still persisted among Quakers a distrust of music, characteristic of the Puritans – and of the Presbyterians of the early nineteenth century. Silvanus had a certain independence of judgment and was willing to question authority. He often attended evensong in York Minster, although he knew that this was viewed with disfavour by the authorities of the school. He managed slowly to moderate the prejudice of his parents and having saved enough to buy a second-hand piano, he taught himself to read music and to play accompaniments. He never achieved any real competence as a pianist but he did have a good baritone singing voice.

Silvanus's father, whose health was now poor following a railway accident, retired from his teaching post in 1874. In the summer vacation of that year, Silvanus made his first visit to the Continent. His French was already fluent but his German was inadequate. He crossed from Hull to Antwerp, then travelled via Brussels to Basle, visiting Luxembourg and Metz on the way. From Basle he journeyed, mainly on foot, through the Juras. During a six-week tour he wrote home often and kept a journal especially for the benefit of his invalid father. He also made dozens of sketches. He describes in terms which express the extravagant enthusiasm of youth his first view of the High Alps seen from the Weissenstein. 'Well,' he wrote, 'one of the dreams of my life has been at last fulfilled. The snow peaks of the Alps are stamped into my mind forever.'

Teaching in a school with the limitations of Bootham was becoming irksome and, coupled with his recognizing the need to obtain better facilities for study, Silvanus resigned at the end of the session to go to London as a full-time student. He had already passed the Intermediate Science Examination and, on the result, had been awarded a bursary to the Royal School of Mines in South Kensington. His subjects were to be chemistry and physics. In October he and another Quaker student, Ernest Westlake, took rooms at 83 Robert Street, South Kensington.

The Professor of Chemistry was Edward Frankland and the Professor of Physics Frederick Guthrie. Of them Silvanus wrote to his father: 'He (Guthrie) is a ponderous Scotchman, and puts in "of course" about thirty times in each lecture. Frankland is a much neater and tidier lecturer. His experimental illustrations are all most excellent.' Within a short time, Frankland had recognized Thompson as a young man of exceptional ability and had begun to take a personal interest in him, advising him to attend lectures at the Royal Institution. Frankland gave him tickets for the Friday Evening Discourses there. The Director of the Royal Institution was at that time John Tyndall. In a letter to his mother, he wrote:

Tyndall's lectures concluded last Saturday. I have made very full notes and embellished them with about eighty scribbles of apparatus, experiments and portraits. They were, as examples of popularised science, admirable: the illustrations brought forward being most excellently adapted to teach the subject . . .

I have learnt a good deal of the method and 'technic' of lecturing by them, and have had the opportunity of seeing what we do not get at all in Guthrie, and very little in Frankland, the swing, the ease, the dash, that makes all the difference between the easy and the tedious lecturer.

The BSc examination was held at the end of October and when the pass list appeared, Thompson's name was one of three who had achieved First Class Honours. He continued to work at the Royal School of Mines on research projects of Professor Guthrie.

University Teacher

At about this time Thompson was admitted a Fellow of the Royal Astronomical Society and he also joined the fairly recently founded Physical Society. It is, however, clear from his letters that the high peak of enjoyment was reached in attending lectures at the Royal Institution. Of T. H. Huxley lecturing *Upon the Comparative Anatomy of the Lower Vertebrates* he wrote: 'The flow of language was perfect and the whole manner most graphic, perspicuous and simple. As a speaker he beats Tyndall hollow.'

Towards the end of January, Guthrie and Thompson carried out a short investigation into an electrical phenomenon reported by Dr Beard of Baltimore as a 'new form of electricity'. By 4 February, Thompson was writing to his father – 'having completed for the time being the investigations in hand, I communicated the substance of them to the Physical Society in a paper which you will find briefly reported in this week's *Athenaeum*. We had a very interesting discussion about the new manifestation afterwards. It is originally an American discovery, but the men over there fancy it to be a new force, while our experiments go to show that it is simply a variety of induced electricity.'

He described the phenomenon as follows:

> When a galvanic current passing through the coils of an electromagnet is rapidly interrupted, minute bright sparks can, at the moment of interruption, be drawn from the electromagnet core by an earth wire or any conducting substance held in the hand and presented to the core.

In a series of well-conceived experiments, Thompson demonstrated that the electromagnet core became electrified by *electrostatic induction*, there being an appreciable electrostatic capacitance between the winding of the electromagnet and the surface of the iron core. In other words, an electrostatic charge was induced on the core by its exposure to the electric field of the surrounding winding. The investigation was subsequently reported in the *Philosophical Magazine* in September 1876, under the title 'On some phenomena of Induced Electric Sparks'. This was Thompson's first research paper.

On Friday 11 February 1876 William Crookes, then President of the Royal Society, gave an evening discourse at the Royal Institution on *The Mechanical Action of Light*. Crookes was demonstrating, with the superb experimental technique of which he was master, his remarkable scientific device, the radiometer.

Structurally, the device was brilliant in its simplicity – four mica vanes fixed radially with their planes vertical at the ends of horizontal wire cross-arms, mounted on a pivoted vertical spindle. The assembly, which could rotate freely in its pivots, was mounted in an evacuated glass bulb about the size of an incandescent electric lamp. The vanes were blackened on one side and bright on the other and were mounted to 'face' in the same sense of rotation. On exposing the radiometer to a source of light the assembly would start to rotate, the speed of rotation increasing with the intensity of the source. Such a device, if it could be made sufficiently sensitive, should rotate, on exposure to light, as a result of the different mechanical reaction to photons reflected from the bright surface and to photons absorbed at the dull surface, as could be deduced from Maxwell's equations of the electromagnetic field. The direction of rotation was *opposite* to that predicted and calculation showed that, for true electromagnetic action, the effect would be far too small to be observed.

Intrigued and excited by the lecture, Thompson seized an opportunity, which soon occurred at the Royal Society, to speak to Crookes on the matter. Immediately afterwards he began to make his own radiometers. Some two years elapsed before the behaviour of the radiometer was satisfactorily explained. Osborne Reynolds conceived a beautiful critical experiment which demonstrated that the phenomenon was due to residual gas molecules, differential heating of the faces of the vanes resulting in higher recoil velocities for molecules from the blackened sides. Reynolds suggested suspending the radiometer by a fine fibre and noting whether reverse rotation of the envelope took place, which it did.

In a closely-knit religious group like the Society of Friends, it was to be expected that the young Thompson would bear letters of introduction to his father's Quaker friends in London. Religious communities throughout the world tend to serve in furnishing places of social resort for their adherents when away from home. For the young, these are regarded as a 'safe refuge'. Thompson began to attend one of the largest Quaker meetings in London, the Westminster Meeting in St Martins Lane, and before long he was a welcome guest in many homes. Alfred W. Bennett, lecturer in Botany at University College and Bedford College, hospitably kept 'open house' for Quaker students.

Thompson's association with the Westminster Meeting had at least two far-reaching consequences. He became a member of a small literary and artistic circle called the Friends' Portfolio Society and he continued throughout his life to attend its meetings. Among the social events of the winter were several lectures. In the session 1875-6, one had been given by A. W. Bennett, presumably on a botanical topic, and another by the distinguished judge, Sir Edward Fry. Thompson was invited to give a lecture on any scientific topic and he chose the subject of *Comets: their purpose in the Universe in the light of recent research*. He went to great pains to illustrate his lecture by diagrams and experimental demonstrations and to present his arguments with cogency and clarity. Among his hearers was a young woman, Jane Henderson, whom he had already met, and who was the eldest daughter of James Henderson, a member of the Quaker congregation. Five years later she was to become his wife.

Just after Easter 1876, Thompson mentioned, in a letter to his father, that he was thinking of applying for the post of Professor of Chemistry in the recently established University College at Bristol. He wrote, 'I saw the advertisements about a fortnight ago, but felt inclined to pass the matter by, as rather beyond my sphere at present. However, a few days ago, Mr Lodge, Professor G. C. Foster's right-hand man, with whom I was having some chat, said: "Why don't you go in for Bristol, it's the very place for you?" '

He consulted Dr Frankland, meeting, for the first time, the all-too-common response in such circumstances, that the professor had already been approached by another candidate. Frankland, however, was encouraging, indicating that Thompson's work was very promising and that, with a little more to show, he would be well in the running for more senior posts. Thompson did apply but was unsuccessful.

Shortly afterwards, a second vacancy in the College was advertised, a lectureship in Physics, and Thompson applied for it. He had already planned a summer vacation in Germany, intending to visit the University of Heidelberg and he set out armed with a letter of introduction from Professor Guthrie to Professor Bunsen. Travelling leisurely up the Rhine, visiting some of the smaller places, Rolandseck, Bacharach and Boppard, he broke his journey at Bonn to visit Doctor Geissler, maker of the famous discharge tubes. He then reached Heidelberg and settled in a *pension*. In the first mail he received from England was a letter offering him the appointment in Bristol.

The picturesque ancient town of Heidelberg, with its unfamiliar manners and customs, its strange combination of formality and casualness in the university, charmed and intrigued him. His letters home, illustrated by little pen-and-ink sketches, underline the deep impression made on him by his 'Summer Semester' there.

A letter written to his father shortly after taking up his post in Bristol suggests that Thompson must have suffered mild parental censure for extravagance. 'The other source of expenditure,' he wrote, 'has been books. While in Heidelberg I spent nearly £2 on French and German books and got some most helpful ones for my work here. On my return here I got some more, including Ganot's Physics, Professor Tait's excellent new book *Recent Advances*, two books on Physical Measurements and a book on Higher Mathematics at which I am taking daily doses. . . . I had purposed, a little while ago, getting Guillemin's two beautiful volumes, *Forces Physique* and *Applications de Physique* but they cost 20s each, so while I was in London I made up my mind not to buy them. I took also, at Kensington, a last fond look at Helmholz's splendid work on the *Sensations of Tone*, a perfectly wonderful book . . . I finished yesterday reading another interesting book that I have got lately – Balfour Stewart on the *Conservation of Energy*, quite a readable book and one teeming with the latest information.'

The verve and enthusiasm with which Thompson's early letters from Bristol are infused testify not only his own dedication but also to the stimulating and, at the same time, congenial atmosphere in which he found himself. Some of his Quaker contacts were influential. He was almost immediately accepted into the circle of the public-spirited, the intellectual and the cultured. In one letter he recounts that shortly before giving his own opening lecture he attended the opening address of Professor Rowley on Modern History. Apparently the professor made a smart attack on the idea that the study of natural science could be a cultural activity. Of this Thompson instantly 'made a note', and recast the end of his address to rebut Professor Rowley's assertion.

Towards the end of the year Thompson's domestic arrangements and his technical assistance in the college were affected by the arrival of his young brother, Jack. There were two younger brothers, Thomas and Jack. Thomas was in training for a business career and Jack had been apprenticed to a pharmacist in Scarborough. Jack's health had, however, been seriously undermined by an attack of pneumonia and it was considered unwise for him to continue his apprenticeship under the rather severe winter conditions of the North East coast. Thompson had offered to help by taking Jack to live with him in Bristol and engaging him in the work of the laboratory and in lecture preparation. In one of his letters he recalls how he and Jack uncovered a 'cache' of scientific apparatus in the storerooms of the Bristol Museum.

On Tuesday we ransacked the garrets of the museum and found an ocean of fine-apparatus. Two glorious air-pumps (one must have cost £60) in good condition but more than an inch deep in dust and quite forgotten. Also 70 cells of Wollaston's battery. A large plate glass electric machine. A battery of 12 Leyden jars, each of $2\frac{1}{2}$ gallon size – and an Attwood's machine worth at least £50 at the present moment, besides a lot of lesser apparatus.

I shall have the opportunity of using all of these things for my Christmas lectures and they are literally 'a find'.

The Christmas Lectures were to be designed 'for a juvenile auditory',

2 *In the Black Forest, by Thompson*

much like the famous series at the Royal Institution in London. Thompson gave six on *The Forces of Nature*, covering sound, light, heat and electricity. The lectures were oustandingly successful and established Thompson's reputation as a good popular lecturer.

Despite his teaching commitments and an increasing list of external lecture engagements, Thompson managed to sustain and extend his research interests. He prepared for the British Association Meeting to be held in Plymouth in the autumn a paper on *Binaural Audition*, a subject which was to hold his attention for a long period ahead.

3 *Silvanus P. Thompson in 1877*

To facilitate his demonstrations of electrical phenomena before large audiences he devised a simple but most effective form of mirror galvanometer and this was the subject of a further communication to the same meeting. Before the autumn came, however, Thompson made a second expedition to the Continent, this time in company with a fellow student from London, Walter Palmer. They travelled up the Rhine, then into

Switzerland, calling at Zürich where they visited the Polytechnic. This institution, which had been founded some twenty-five years earlier, brought from him the comment that: 'There is here the most complete and excellent Engineering School in Switzerland – probably in the world'.

Thompson and Palmer walked over the two Sheideck Passes and the Grimsel, then up to Zermatt and finally crossed into Italy by the Theodule Pass. Here they had as a guide, Peter Taugwalder. He was one of the three survivors of the ill-fated Whymper expedition which in 1865 made the first ascent of the Matterhorn, on which tragically four of the party were lost in a disaster on the descent.

Thompson's opening address of the new session at college was entitled *The Methods of Physical Science*. In the course of outlining the historical development of the scientific method he made reference to the experimental researches of William Gilbert, physician to Elizabeth I and James I, and to his famous book, *De Magnete*. Describing magnetic phenomena and regarding the earth as a great magnet, this was the first major scientific work to be published in England.

In retrospect, there are in the record of Thompson's first year in Bristol some of the strands which predominated in the whole pattern of his later life.

The Young Professor

By the spring of 1878 Silvanus Thompson felt able to present himself for the examinations for the degree of DSc in the University of London. He had chosen the branches of Optics, Heat and Sound. The examination was wholly by written papers, one in each subject, and the examiners in that year were Professors A. W. Reinhold and Balfour Stewart. The examination was held in June and his success was duly announced in the pass list.

Sometime in the autumn, Thompson received an advancement he had not sought. The College had created, with a very modest endowment, a Chair of Physics, and he was elected to it. Thus, at the age of 27 he assumed superior academic responsibility. The distinction of the Professorship was, however, more in the title than in the substance, and he discovered that the Principal, Professor Alfred Marshall, a distinguished economist, and the members of the College Council did not appreciate the needs of a scientific department for technical assistance and equipment. His teaching and the illustration of his lectures by effective demonstrations gave Thompson immense satisfaction, but he found he had to construct a good deal of his own apparatus. Lack of resources did not, however, inhibit his pursuit of research. Intellectually curious, as any first-class teacher must be, the very work of teaching threw up a sequence of questions meriting investigation, and these, in due course, resulted in short scientific contributions.

Two communications were made to Section A of the British Association at the annual meeting held in 1878 in Dublin. One of these reported an investigation which had been in progress for some time in collaboration with Oliver Lodge, then a young assistant to Professor Carey Foster at University College London. This concerned the unilateral electrical conductivity of a crystal of tourmaline. It was considered that the observed effect was analogous to the phenomenon discovered by Kelvin, then Sir William Thomson, in which a thermal gradient in a conductor is associated with a potential gradient. The second communication reported further work on binaural audition.

Jane Henderson attended the Dublin meeting as a young reporter on the literary staff of the Glasgow *Daily Mail*.

With his family background of Quaker idealism, the young Professor Thompson felt it a social obligation to devote something of his ability and influence towards ameliorating the condition of the working classes. At the hospitable home of Dr John Hall Gladstone, Thompson had spent many a happy Sunday evening during his student days in London and it may well have been there that his interest in the social influence of education was first seriously aroused. He had kept in touch with Gladstone and corresponded with him regularly on educational matters. Gladstone was a member of the London School Board. During his service as Chairman, he had been successful in introducing a more systematic teaching of handwork and of simple nature study into the curriculum of schools. He was an early campaigner for the introduction of more practical studies in schools and a reduction in the emphasis on the scholastic content of the curriculum. He was a chemist and viewed the problem of education from the standpoint of a scientist. In this, he and Silvanus Thompson were at one.

Thompson had already realized that technical education in England lagged seriously behind that in several Continental countries and that scientific advance would increasingly produce demands for new technical skills which would not be met. Determined to assess the matter competently, he devoted several vacations to studying the provisions for technical education in France, Germany and Switzerland. His first public comment on it was made at Cheltenham, at a 'Social Science Congress' in 1878. He gave a paper, subsequently published as a pamphlet, entitled *Technical Education. Where should it be given?* The pamphlet was brought to the attention of John Ruskin, who wrote with qualified appreciation, asserting forcefully, in relation to the decorative arts, that 'No good decorative work ever has been done in England – or ever will be'.

Further visits to the Continent gave Thompson a comprehensive picture of the systems in operation in the *Gewerbeschule* and *Polytechnicums* of Germany and Switzerland, and in the *Écoles d'Apprentis* of France. During the summer of 1879 he wrote and published a booklet entitled *Apprenticeship Schools in France*. The well-regulated and efficient, even if somewhat regimented, character of the Continental systems must have appealed to him, perhaps more so because there was no *system* of technical education in England. It was, however, no narrowly conceived project for the improvement of technical education that was in Thompson's mind, but a wide-ranging concept by which his imagination had been seized. He conceived that science and, in particular, electrical science would become the most powerful driving force, in a physical sense, in social and industrial development. He recognized technical education as the means by which scientific knowledge would be translated into action. He saw technical education as a unity embracing the acquisition of craft skills, the association of elementary scientific knowledge with craft practices, the inculcation of scientific principles at what would now be termed technician level and, finally, the production of the engineer-scientist or technologist.

In a very real sense Thompson devoted his life to bringing his great concept of technical education to practical realization. He pursued his studies of the matter both at home and overseas. Following a tour, during the long vacation of 1879, of Trades Schools in Yorkshire, the North of England, and in Glasgow, where he stayed with James Henderson, by then Superintending Inspector of Factories, he went to Germany. There he visited the great Technical Institute at Charlottenburg. Going on to Hanover and Chemnitz he reached Leipzig where he developed an attack of laryngitis so severe that he sent an appeal for help to his youngest brother, who came and stayed with him until he was well enough to travel home. He was unable to fulfil some of his autumn engagements and was left with his throat permanently weakened.

In December 1879, he read a paper to the Society of Arts entitled *Apprenticeship, Scientific and Unscientific*. It was an important occasion, giving Thompson the opportunity to discuss in some detail the deficiencies of the provisions for technical education then existing in England, to describe his own ideas as to the essential requirements and to give his impressions of technical schools abroad. In the course of the lecture he advocated the appointment of a Minister of Education. Thomas Henry Huxley, the great protagonist of Darwinism and at that time Professor of Zoology at the Royal College of Science, was in the chair. In the discussion which followed, Professors Ayrton and Perry, the former Professor of Electrical Engineering and the latter, Professor of Mathematics at the Finsbury Technical College, expressed the opinion that England was too conservative to make use of trade schools and that the methods adopted in Continental countries would not be suitable for this country. Huxley, who was clearly much in sympathy with Thompson, summed up at the end, referring to the City & Guilds of London which were established to aid their respective trades, and declared that, if the people of this country did not insist on their wealth being applied to its proper purpose, they deserved to be taxed down to their shoes.

Early in the following year, 1880, Thompson made another journey to Glasgow to stay with the Hendersons. The object of his journey, he had already hinted to his mother, was to become engaged to Jane Henderson. Theirs was a friendship which had matured from his student days in London in the congenial atmosphere of a quiet, cultured, Quaker family similar to his own.

Jane Henderson and Silvanus Thompson were married on 30 March 1881, in the old Friends Meeting House in Glasgow.

Evidently, at about the time of his engagement, Thompson had been becoming increasingly dissatisfied with conditions in the College. The lack of effective financial support in building up his own department, and indeed in developing the College as a whole, was so discouraging that he began to look around for other posts, but none of his applications were successful.

Perhaps as an antidote to frustration, he bought a house in Clifton in which to settle down to write a text book. In November 1879 Thompson had written to Macmillan & Co., the publishers, offering to write a textbook of physics, or alternatively, an elementary treatise on electricity and magnetism. Macmillan replied cautiously:

'... On the whole, though fully recognizing the importance of such a book, we are inclined to pause before committing ourselves, at least at present, to a treatise on so great a subject as Physics from a writer of whom we have had no experience. You will, we feel sure, excuse our saying frankly that, although all we hear of you is most hopeful for the future, we do not feel able yet to regard you as certain to turn out what would be a standard text-book of the subject.

The second, as being a smaller venture, we are disposed to view more favourably, an elementary treatise, that is, on Electricity and Magnetism, in our Class Book series. We should be glad to know what space....'

The second book was to become his *Elementary Lessons in Electricity and Magnetism*. In Thompson's lifetime the book ran through more than forty editions and reprints. It established his reputation as an outstanding expositor of scientific principles and it carried his name throughout the world.

Despite the professional frustrations, Bristol proved to be a very pleasant place, socially and intellectually, for the young professor and his wife. One of the great events of that year was the Jubilee meeting of the British Association in York. Sir John Lubbock, the mathematician, was President. He had been first Vice-Chancellor of the University of London and treasurer of the Great Exhibition of 1851. William Thomson (later Lord Kelvin) was President of Section A, the Mathematics and Physics Section. Hooker was President of the Geographical Section. Huxley and

Darwin were present. The *soirée* was held in the Art Gallery which was illuminated by incandescent electric lamps.

Thompson read four papers, one on Volta Electric Inversion, one on the Opacity of Tourmaline, the third on A New Polarising Prism, and the fourth a further communication on Binaural Audition.

Author and Public Lecturer

Silvanus Thompson spent eight years in Bristol, two as lecturer and six as Professor of Physics. His scientific interests, as evidenced by his papers published during this period, ranged widely, but even more significant was his increasing involvement in propagating a knowledge of science throughout the general body of the community.

It was his interest in the dissemination of scientific knowledge that prompted him to apply, in 1880, for the newly created post of Organizing Director and Secretary of the City & Guilds of London Institute. Philip Magnus however was appointed to the post and five years were to elapse before Thompson became directly involved in the work of the Institute, when he took up the appointment of Principal of the Finsbury Technical College.

Technical education had been struggling for recognition from the 1830s. Mechanics' Institutes had been established but little effective coordination had been achieved except in the Manchester area. The attention of Government had been called to the urgent need to facilitate the application of scientific discoveries in manufacturing industry but action had not yet been initiated when, in 1876, a momentous decision was taken by the Guilds of the City of London.

At a meeting on 30 July with the Corporation of the City of London held under the Chairmanship of the Lord Mayor, a resolution was adopted in the following terms:

> That it is desirable that the attention of the Livery Companies be directed to the promotion of Education, not only in the Metropolis but throughout the country, and especially to Technical Education, with the view of educating young artisans and others in the scientific and artistic branches of their trades.

What could be more appropriate than that bodies whose foundation and whose wealth had originated from the great mediaeval Craft Guilds should, in the nineteenth century, foster and support the development of the new technology?

The resolution led to the foundation in 1878 of The City & Guilds of London Institute for the Advancement of Technical Education.

Finsbury Technical College was a teaching institution created by the City & Guilds Institute and it was as Principal of the College and Professor of Physics that Silvanus Thompson was to devote thirty years to the advancement of technical education.

While the years in Bristol had proved to be frustrating and disappointing, in the sense that all the strenuous efforts Thompson made to obtain much-needed financial support for the College failed, they did provide him with a wonderful series of opportunities for giving public lectures. Some were given in Bristol, others in or near London, others, again, in the Midlands, Wales and the North of England. Sound, optics, heat and colour featured in his repertoire but the major topics were related to electricity. The electric light was a subject of great popular interest. To the Royal Society of Arts he lectured in 1881 upon *The Storage of Electricity*. The opening paragraphs of this lecture are worth quoting in full.

> Science has, of late, made two advances, the ultimate importance of which it would be difficult to overestimate. Not many months before he was seized with the mortal illness which robbed us too soon of his rare and unique genius, Professor Clerk Maxwell was asked by a distinguished living man of science, what was the greatest scientific discovery of the last twenty-five years. His reply was 'That the Gramme Machine is reversible'. His far reaching and philosophic mind had perceived that in this phenomenon, which to so many had seemed little more than a curious scientific experiment, lay the principles which, if rightly developed, would render possible the electric transmission of power and, in the solution of this practical problem, bring about social and economic changes the importance of which but few of us here even yet begin to realise.
>
> If we could tonight summon up the noble spirit of the philosopher and ask him to tell us what recent scientific discovery came next in importance to this, I think we should receive the reply 'that a voltaic battery is reversible'. The reversibility in action of the voltaic cell is the counterpart and complement of the reversibility of the Gramme machine, for while the one has solved for us the problem of the *electric transmission of power* the other has solved for us the problem of the *electric storage of energy*.

Thompson was, of course, referring to the discoveries of Gaston Planté, which had originated the lead-acid secondary cell, and to the notable improvements made subsequently by Camille Faure. These advances were almost contemporary with Swan's and Edison's successful development of the carbon filament lamp. That Thompson was over confident in describing the problem as being solved is, nowadays, abundantly evident but, from the viewpoint of the time, the assertion was justified.

A conference was held in Paris in 1881 to consider questions of terminology in relation to electrical science and Thompson was a delegate. There he met Gaston Planté. In the autumn of that year the Crystal Palace Company began to organize an International Exhibition of Electricity, and early in 1882 Thompson received a letter inviting him to give, in connection with the exhibition, four or six elementary lectures on Electricity with special reference to electric light.

In due course he gave four evening lectures, to the evident satisfaction of large audiences. The reviewers and newspaper reporters were not, however, wholly complimentary. Whilst highly commending the basic material, the admirable demonstrations and the skilful presentation, there was criticism of the inclusion of speculative observations about electric charges representing condensations or rarefactions of the ether. In addition there was an imputation of some measure of bias in favour of Edison against Swan in the matter of the development of the incandescent lamp.

In the first of the lectures Thompson forcefully condemned some of the very dubious practices associated with electrical equipment which claimed to produce beneficial medical treatment. He said that, 'the mistake of confounding physiological with medical or remedial effects led to the gross impositions of the quacks and rogues who deal in so-called magnetic appliances and disgrace alike the sciences of electricity and medicine, while knowing nothing of either'. Such a remark could not be expected to pass unchallenged. No doubt he spoke with the buoyant confidence of youth, but the Quaker qualities of forthright expression of opinion and incapacity for compromise were already evident and were to characterize his actions thoughout his life.

The pioneer installations for public electricity supply were Edison's Holborn Viaduct Station in London and Pearl Street Station in New York which were constructed in 1883. In England the Government had in 1879 set up a Select Committee under the chairmanship of Sir Lyon Playfair to consider the position of public authorities in relation to electricity supply and to study the conditions under which private companies should be permitted to provide a supply. The outcome of the Committee's deliberations was the passage through Parliament in August 1882 of the first Electric Lighting Act. Thompson was involved to the extent of preparing a Report and General Advice on Draft Provisional Orders. The Act, as finally drafted, contained a clause which was to prove a most serious hindrance to the development of electricity supply. This was the provision under which municipalities could acquire an electricity undertaking after twenty-one years. This so discouraged investment as almost to inhibit development. Some two years later Thompson was appointed a member of a committee set up to frame clauses in amendments to the Act. It was, however, not until 1888 that the period was extended to forty-two years.

The financial position of Bristol University College was becoming progressively less satisfactory. Government support was still not forthcoming and private benefactions were small. By 1883 finances were so straitened that the College Council had seriously to consider the possibility of making a cut in the salaries of all staff. Another possibility contemplated was that of shutting down completely the arts side of the

College. Thompson was severely critical of the College Council which he regarded as wholly lacking in initiative. These dangerous suggestions moved him to the most strenuous efforts both in correspondence with influential friends and in the press to draw attention to the shameful neglect of a highly commendable and promising educational institution. These endeavours did not bear fruit in his time in Bristol for he was soon to move to London.

Thompson's researches into binaural audition and the intense interest aroused by Graham Bell's recent epoch-making telephone invention combined to turn his attention to possible ways of developing the telephone. He had come across a reference to the work of Phillip Reis during one of his earlier visits to Germany. Then he met, in Manchester, a former pupil of Reis, and he determined to investigate Reis's invention further. In the summer vacation of 1883, he visited Germany and met Reis's widow and son. His investigations convinced him that Reis was, in fact, the first man to transmit audible sounds by electrical means and that he was, in consequence, the real inventor of the telephone. This led him to write a little book entitled *Life of Phillip Reis*, and this was published in 1883. The telephone stimulated Thompson to further invention, to

engage him for the first time as an expert witness in a patent action and to lead him into a commercial venture, 'The New Telephone Company', to which further reference will be made.

In 1882 Thompson gave to the Royal Society of Arts the annual series of lectures known as the Cantor Lectures, on the subject of Dynamo Electric Machinery. These lectures formed the basis of one of his most famous books published two years later.

Dynamo Electric Machinery, a much larger work than *Elementary Lessons in Electricity and Magnetism*, appeared in 1884. It became one of the classics of electrical engineering and has been reprinted several times.

A study of electro-magnetic machines was, for Thompson, a natural outcome of his interest in electromagnetism, and of his profound admiration for Michael Faraday and his electrical researches. While dealing adequately with the underlying physics the book was directed to the technology of machines, giving full and detailed treatment to essentially engineering considerations. Thus, Thompson, basically a schoolmaster and a physicist, had apparently acquired sufficient engineering background to enable him to examine critically the design and performance of the machines of the leading manufacturers of the day. At that time the

4-6
These electrical machine models were used by Thompson to illustrate his lectures

artificial distinction between the engineer and the scientist which was to develop during the earlier years of the present century, had not yet become evident. By 1882 Thompson's interests in electricity had developed markedly on the applied side. In that year, proposed by Professor Carey Foster, he was elected a member of the Society of Telegraph Engineers and Electricians (now the Institution of Electrical Engineers).

One of the outstanding events of the year 1884 for the British Association, and also for Thompson himself, was the annual meeting. It was held in Montreal. This was the first British Association annual meeting ever to be held overseas, and attendance provided Thompson with his first chance to visit the American continent. The President of the Association was Lord Rayleigh. Among the party of distinguished British scientists present were Sir William Thomson, William Ramsay and Oliver Lodge. In a letter to his wife, Thompson mentions the friendly contact he established with Alexander Graham Bell, of telephone fame. Thompson and Oliver Lodge went on from Montreal to Philadelphia via Toronto and Niagara Falls, to attend the American Association for the Advancement of Science, which was meeting soon after the British Association meeting.

7 *Thompson (centre, front row) with students and staff of Finsbury Technical College Electrical Engineering Department, at the beginning of the present century. Sixth from right in the second row is Frederick Handley Page, later Sir Frederick, a leading pioneer of flying and aircraft development*

The Montreal meeting had, oddly enough, an incidental beneficial reaction upon University College in Bristol, and indeed upon other university colleges. William Ramsay, appointed Professor of Chemistry in Bristol in 1880, had in 1881 become Principal of the College in succession to Alfred Marshall. Later, Ramsay achieved great distinction as Professor of Chemistry at University College, London, becoming best known for his discovery of the inert gases argon, helium, neon, krypton and xenon. Ramsay used the opportunity afforded by the journey undertaken after the meeting to discuss with his brother principals the financial plight of the colleges. A collective strategy for a united attack upon Government was worked out and this, in due course, broke down official resistance and obtained substantial financial support.

Finsbury Technical College
Meantime, college affairs in Bristol were, in Thompson's own words, 'drifting from bad to worse'. The scheme for a reduction in salaries was put into effect and the suggestion that the arts side of the College should be discontinued was raised again. At this stage the opportunity occurred for Thompson again to seek an appointment in London, under the City & Guilds Institute. Philip Magnus, the Directing Secretary, wrote letting him know that the post of Principal of the Finsbury Technical College would fall vacant when Professor W. E. Ayrton took up his appointment to the Chair of Electrical Engineering at the Central Technical College.

The City & Guilds of London Institute took the whole field of applied science for its province, comprising the most advanced technology, extending into what would now be termed the technician field and spreading out to areas of quite specialized craftsmanship. Two major teaching institutions were established. Finsbury became the prototype of the English technical college and of the great London polytechnics, while the Central Technical College in South Kensington developed at the equivalent of university undergraduate and postgraduate levels. Thus, Finsbury was conceived as covering a much wider range of teaching activity than the Central institution. Building commenced at Finsbury in 1880 but teaching had been initiated on a small scale in November 1879 following the appointment of Professor H. E. Armstrong (Chemistry) and Professor W. E. Ayrton (Electrical Engineering). Temporary accommodation had been arranged in a school adjacent to the site of the new college buildings which were opened in 1882.

Thompson and his work were already well known to many members of the City & Guilds Institute. He had, in fact, drawn up, at the invitation of the Committee, a scheme for the organization of the Central Technical College. He was one of seventeen applicants for the post. Another was his colleague and principal, William Ramsay. Among those who supported Thompson's candidature were Alfred Marshall, now Professor of Political Economy in the University of Cambridge, and Dr Benjamin Jowett,

Master of Balliol College, Oxford, the great scholar about whom the humorously malicious lines were written, 'I am Master of this College. What I don't know isn't knowledge.'

Thompson was appointed to the post early in March 1885. He moved almost immediately to London to take up his duties at Finsbury after the Easter vacation. The Council of the Bristol College was naturally anxious that he should complete his teaching for the session and this he was able to do. Shortly after Thompson's arrival at Finsbury, he was joined by Raphael Mendola as Professor of Chemistry, in succession to H. E. Armstrong who, like Ayrton, was moving to the Central Institution. Mendola and Thompson had met and become friendly during the latter's first term at the School of Mines. This was for Thompson probably the happiest appointment to his staff, for Mendola was the most congenial of colleagues. He was an able chemist and he shared with Thompson many cultural interests, in particular, natural history and literature.

It was not without regret that the Thompson family – there were now four daughters – left their congenial circle of friends in Bristol and the delightful surroundings of their house in Clifton. They had enjoyed to the full living close to the Downs and within a short walk of the cliffs overlooking the Avon Gorge, with the delight of the daily observation of nature lying almost immediately at hand. The initial choice of an area in London in which to live rested on it being near to some old friends. They found a house in Bayswater 'in a dingy row of high houses called Arundel Gardens, the attraction to the house being the large square behind where the children could play'. Dr Gladstone, William Crookes and Professor Ayrton lived only short distances away.

Silvanus Thompson's life was now centred on Finsbury and remained so up to his death thirty years later.

As set out in the prospectus, the objects of the college were the education of

I Persons of either sex who wish to receive a scientific and practical preparatory training for intermediate posts (as for instance foremen or managers) in industrial works.
II Apprentices, journeymen and foremen who are engaged in the daytime and who desire to receive supplementary instruction in the art and practice, and in the theory and principles of science connected with the industry in which they are engaged.
III Pupils from middle class and other schools who are preparing for the higher scientific and technical courses of instruction to be pursued at the Central Institution.

The College therefore fulfilled the function of a technical finishing school for those entering industrial life at a comparatively early age; of a supplemental school for those already engaged in factory and workshop;

and of a preparatory school for the Central Institution. Full-time courses were later developed on the basis of either a two-year course leading to the College Certificate or a three-year course for the successful completion of which the College Diploma was awarded.

The objects of these courses were described as follows:

> The courses of instruction and the practical work in the laboratories and workshops are arranged solely with the object of providing a training on broad, practical and scientific principles, in those matters which experience has shewn best to fit students to enter some one of the many careers connected with Civil, Mechanical, Electrical or Chemical Industries or professions.

It is of interest to note that, in a report to the Governors in 1901, Thompson strongly advocated the setting up of what would now be termed 'Sandwich Courses'. It does not appear however that the recommendation was ever implemented.

In addition to the major departments of the college which were all associated in one way or another with industry, there was a large Art Department.

During Thompson's first year in office there were 156 day students and 912 evening students. At the end of the year, in a critical review of the operation of the college, he wrote:

> '... the education given in the college presents several points in marked contrast to an ordinary college education. The laboratory, the workshop and the drawing office take up the main portion of the students' time. For every hour in which the student is being talked to in the lecture room, there are two hours in which he is instructing himself by actual work.
>
> Textbooks are almost unknown; the students acquire their facts and draw their inferences not from books, nor from the *ipse dixit* of the teacher, but from the things themselves. The results of this scheme of instruction are briefly this; that the students who have followed out their course enter industrial life under much more favourable conditions than otherwise they could have done. They pick up in the shops, in two or three years, more than they could have done in five or six years under the old apprenticeship system...'

Thompson considered a good technique of taking notes on lecture courses and of recording laboratory observations to be of the greatest value. Indeed, so highly did he rate the importance of this matter that he would hold a 'note-taking' class at the beginning of a new session and would give up part of a Saturday morning to 'drilling' students in the art.

Thompson's attitude to work was uncompromising. He 'was very strict with first-year students; if they wasted their time, were indolent and unpunctual, showed no purpose in their work, and failed in the first year's examinations, he would frequently advise their parents or guardians to take them from the college and put them to other work. In a very few cases they were allowed to repeat the first year course.'

In accordance with the practice of the time the physical facilities of the college were designed exclusively for work. There was no refectory nor were there any student common rooms. When, in the early days, a *soirée* was held for students and their friends, the party would overflow into rooms kindly lent for the occasion by the adjacent Cowper Street School.

Despite what would be regarded nowadays as almost insuperable barriers to the establishment of a corporate spirit, the college slowly built up a quite astonishing esprit de corps. Thompson's gospel of hard work was infused with a certain gaiety of spirit and pride in achievement and he made the welfare of every student – using that term in its full sense – his personal concern. He must have recruited to his staff men of like mind and similar enthusiasm, for it is doubtful whether any other institution ever engendered greater loyalty and affection in its Old Students.

Life in London as Principal of a new and developing technical college was at once harassing, exacting, stimulating and rewarding. In addition to being Principal of the College, Thompson was Head of the Department of Physics, and for certain periods he delivered as many as ten lectures a week, some being given to evening classes. The Chairman of the Finsbury Governors for the major part of this term of office was Lord Halsbury.

There is no indication that Thompson was on anything other than the friendliest terms with his Governing Body. His terms of appointment permitted him to undertake consulting work, but he was required to seek the Governors' approval specifically for each commission. This he found somewhat irksome but the Governors delayed a considerable time before giving a 'blanket' permission. Funds for the purchase of equipment must have been inadequate. A manuscript draft list of equipment prepared for submission to the governors in 1892 contains two revealing marginal notes. One relating to telephone equipment read, 'The only telephone apparatus we possess is two single receivers to hold to the ear. We have no transmitter at all and cannot even shew our students how a telephone set should be erected or used.' The other relating to a motor-alternator set states, 'We sorely need this for generating alternate currents. The only alternate current dynamo we possess is an old Siemens machine built in 1877 which once lay three months at the bottom of Sydney Harbour (in the capsized *Austral*) which we bought for £15.' The *Austral* was a passenger vessel belonging to the Orient Steam Navigation Company which, in November 1882, had completed her second voyage to Sydney. Due, apparently, to a quite incredible lack of supervision, she sank at her moorings while she was being coaled. She was refloated and sailed back to England with, presumably, the alternator still in her hold.

Administration notwithstanding, Thompson managed to maintain some research interests and he pursued most actively the development of new lecture demonstrations to illustrate his teaching and his public lectures. He was, for example, greatly interested in the subject of electroplating. He devised a method of electro-depositing cobalt so as to give a beautiful untarnishable silver-grey surface useful for decorative purposes. His method was the subject of a Royal Society paper in 1887 and of a patent.

Life in London presented almost too many attractive prospects for a man of Thompson's perceptive mind and wide-ranging sympathies and interests. One of his earliest actions was to become a member of the Royal Institution. For one with so deep a veneration for Faraday, so keen an awareness of the social influence of science, and who had, during his previous residence in London, some experience of the Royal Institution's unique attractions, membership could not be taken up too soon. As a centre of scientific research and perhaps, even more importantly, as a social meeting place for men of science, their families and their friends, the Royal Institution had achieved a remarkably influential position in middle-class society.

The Christmas Lectures for children, today so widely publicized on television, were eagerly awaited in those times, as one of the major events of the winter for young people. The Friday Evening Discourses given always by distinguished speakers and presented with full traditional ceremonial, were often quite brilliant functions. That the occasion of the first Friday Evening Discourse, attended in full membership by the Thompsons, was marked as a red letter day is obvious from a note written by Mrs Thompson following 22 January 1886, giving her impressions of the beautiful building, the social scene, and recording the anticlimax of Professor Tyndall's address on *Wave Forms*. She wrote 'Poor old man, he maundered on for an hour and twenty minutes, repeating himself over and over'.

During the next thirty years the Thompsons were to be regular attenders on Friday evenings at the Royal Institution and Silvanus was himself to deliver both Discourses and Christmas Lectures.

In 1890 Thompson was elected to membership of 'The Sette of Odd Volumes', an esoteric and exclusive literary club founded in 1873 by Bernard Quaritch, the London bookseller. The Sette comprised a wide cross-section of interests in art, literature, the law and science. Dining once a month, the Sette concealed behind a façade of quaint conceit and innocent drollery a valuable interchange of knowledge and opinion, and the printed proceedings furnished contributions of high literary merit. Thompson was elected to the Sette as 'Brother Magnetizer'. In 1904 he achieved his 'Oddship', that is to say, the Presidential Chair. For their dinners the Sette prepared elaborate menus and programmes wittily illustrated by poems, apt quotations and charming caricatures. Clearly Thompson's scholarship and sense of fun reacted to the atmosphere of the Sette with the utmost enjoyment and stimulated him to make noteworthy contributions and to scintillate in discussion. Inevitably, as Principal of one of the Colleges of the City & Guilds Institute, Thompson was a guest at the dinners of the great City Companies. His membership of the Society of Telegraph Engineers and of the Physical Society involved him in scientific meetings and in social functions. His competence and skill as a speaker both in scientific debate and on social occasions were quickly recognized and he became much in demand as an after-dinner speaker. It is noteworthy that Thompson never deviated from his principle of total abstinence from alcohol.

In 1891 Silvanus Thompson achieved the supreme scientific distinction of election to the Fellowship of the Royal Society. In the citation he is described as 'Distinguished for his acquaintance with the science of electricity, more particularly in its experimental and technical aspects'.

It was about this time that the family moved from Bayswater to the less foggy and much more attractive area of West Hampstead, a suburb then almost on the outskirts of London. A spacious house standing in its own garden was given the name *Morland* after the Thompson ancestral home in Westmorland. In this quiet home the family of four daughters grew up. Here Thompson produced his astonishing output of published work, and

8 *A musical evening at the Thompson home. S.P.T. with his wife and four daughters*

9   *Thompson's bookplate. His library, which contains many rare books, is now in the possession of the Institution of Electrical Engineers*

welcoming hospitality was extended not only to personal friends but to many of the leading scientists of the day.

Thompson's increasing involvement in the work of the International Electrical Commission and his membership of a number of European scientific societies resulted in visits to the Continent several times in most years. Switzerland was the favourite venue for summer holidays. These visits provided the opportunity for him to exercise his talent for languages for he became thoroughly fluent in German and Italian, as well as French. He would use his skill to good effect in making contributions to conferences in the local language whenever the opportunity offered. In a deeper sense he valued his international contacts for their influence in diminishing, even if in only a small and personal way, that international mistrust which he recognized as one of the greatest social dangers of the time.

That Thompson's interests were literary as well as scientific accounts for his building up a splendid private library. *Morland* had a basement billiard room opening by a large low window on to stone steps leading up to the garden. This became the library and Thompson's workroom.

Bookcases lined the whole wall space but tables, cabinets and even window ledges carried piles of periodicals, together with portfolios into which were sorted papers relating to books being written or revised, calculations

10 *Two Christmas cards sent to relatives and friends by Professor and Mrs Thompson, 1893 and 1903*

for practical work in the college or possibly documents relating to some patent action. There was strict order in the apparent chaos which was never to be disturbed by the duster of the housemaid. Thompson's books were his friends. Many of them he would annotate with marginal comments. He had a book plate of his own design, with the family coat of arms as its main feature. In the course of his lifetime, he collected many rare books. Such books related chiefly to great men with whom he was dealing biographically. Perhaps his greatest treasures were copies of William Gilbert's treatise on the Magnet, of which, ultimately, he possessed five.

Thompson and his friend Conrad Cooke initiated the celebration marking the tercentenary of Gilbert's publication of *De Magnete* in 1600. Thompson's library, now one of the great ornaments of the Institution of Electrical Engineers, was purchased after his death by the Institution as a memorial to his distinguished service to the profession. One of the largest files in the library may well have been the one which he accumulated in preparing his great biography of Kelvin. Another would be that relating to the University of London. As a London graduate, he had become a member of Convocation and he took very seriously his filial duty to his Alma Mater. The question of the reform of the University was a very live issue when the Thompsons moved to London. The reform under consideration was that of the promotion of a teaching university to replace, or at least to modify, the purely examining body which the University of London had been since its foundation in 1836. Thompson campaigned for the reform but he was critical of the scheme by which it was proposed to implement it. He found the politics associated with the matter distasteful and although elected to the Senate in 1900 he became sadly disillusioned.

As a Quaker and a pacifist, Thompson had to dissociate himself from some of the movements which he regarded as militaristic, at the time of the Boer War. From his wide-ranging European contacts he felt increasing anxiety in the early years of the century about the prospects for the maintenance of world peace. When, in 1914, in Lord Grey's words, the lamps were going out all over Europe, the light of Thompson's gaiety of spirit was dimmed. With the courage engendered by his faith, he carried on, always hopefully, until in 1916, still in harness, death overtook him.

Research, Royal Institution Lectures

While Thompson's wide range of interests could undoubtedly be matched by that of many of the cultured men of his day, few could equal him for contributing so notably throughout so wide an area. He left his mark on science, technology and education, literature and art, and on sociology and religion.

In the field of pure science Thompson's work was chiefly in electricity, magnetism and optics. It has been described as contributing to the 'discovery of new experimental illustrations of the physical laws'. He gave about half a dozen papers to the Proceedings of the Royal Society, and upwards of a hundred papers to the Physical Society, the *Philosophical Magazine* and *The Electrician*. Typical of his novel demonstrations of physical laws was his illustration of the *Magnetic Action of Displacement Currents in a Dielectric* described in a short paper under that title in the Proceedings of the Royal Society in 1889. Some twenty years earlier, Clerk Maxwell had developed his electromagnetic theory, on the basis of his fundamental postulate that the changing electric induction accompanying the charging of a capacitor had, associated with it, a magnetic field corresponding to that of a conduction current. At the time of Thompson's paper the phenomenon was still little understood. Thompson embedded, in the mid-plane of a slab of paraffin wax a toroidal coil of fine insulated copper wire wound over a ferromagnetic core, presumably of iron wire. Connections to the coil were brought out to a head telephone. He then applied tinfoil to the upper and lower faces of the slab, thereby forming a parallel plate capacitor. When the capacitor was excited by an induction coil, noise was heard in the telephone indicating the presence of varying magnetic induction in the core. Subsequent consideration of this matter has shown that this demonstration is necessarily inconclusive, for the magnetic field may be regarded as arising *either* from the displacement current or from the conduction currents flowing in the electrodes and the connecting wires.

While no specially noteworthy advance in fundamental knowledge is associated with his name, many important detailed observations were made by him. For example, his experiments on binaural audition, reported in three papers to the *Philosophical Magazine* in the years 1877, 1878 and 1881, gave information which is quite basic to the present-day practice of stereophonic sound reproduction.

In his paper of 1876 to the British Association he reported that, with tuning forks producing beats, the beats were still heard if the notes were conveyed separately to each ear. Then in 1878 he utilized the very recently invented telephone to investigate the effects of displacement of phase, pointing out that differences in path length to two ears will give rise to such displacements.

He further noted that when two simple tones of equal intensity in unison reached the separate ears in opposite phase, the resulting sensation was that of the sound being located on the median plane at the back of the head. When the tone intensities were unequal, the acoustic 'image' moved from the median plane towards the ear in which the sound was louder.

In his 1881 paper, Thompson deals with the controversial question of whether beats and difference tones are purely subjective or have objective reality. 'The beats,' he says, 'are objective; they can be seen in a manometric flame if the primary tones are sufficiently loud. The difference tones in the mistuned consonances in question are also objective in as much as any suitable asymmetrical resonator will reinforce them. What is the ear but a complex resonator . . . and, if by reason of asymmetry in the ear the higher terms of the displacements can generate resultant displacements whose periods are those of the tones in question and force them upon the sentient apparatus, then the same tones can . . . be taken up by any other suitable asymmetrical resonator'.

In optics he devised a method for the measurement of the optical constants of lens combinations of short focal length. His main interest was, however, in polarization phenomena, and his work on the design and manufacture of polarizing prisms made a substantial advance in practice in this field.

His special interest in the optical polarization produced in crystals of tourmaline, which arose during his early association with Oliver Lodge, resulted in a paper communicated to the Physical Society in 1881 and published subsequently in the *Philosophical Magazine*.

Doubly refracting crystals such as tourmaline resolve incident light into two components polarized in perpendicular directions and travelling with different velocities.

This optical anisotropy might, on Maxwell's hypothesis that light is an electromagnetic disturbance, result from a corresponding anisotropy in the electrical properties of the crystal, namely the dielectric constant, the magnetic permeability and the resistivity.

'If it can be shown,' wrote Thompson, that Tourmaline crystals are better conductors of electricity in one direction than another it can be deduced, as a consequence of Maxwell's theory, that they will be better transmitters of light in one direction than another and that they will absorb more of those rays of light whose vibrations, consisting of electric displacements, lie in the direction of the best electric conductivity.'

Thompson developed the appropriate solutions of Maxwell's equations. He then examined critically the various possibilities and concluded that the phenomenon could most probably be ascribed to a difference in electrical conductivity along the optical axis from that in a plane normal to it. There were, however, no measurements available against which the hypothesis could be tested.

Among the many subjects upon which Thompson lectured and carried out research, magnetism takes a prominent place. His interests ranged from strictly scientific studies of the characteristics of electro-magnets and permanent magnets, related to their structure and metallurgical composition, to terrestrial magnetism and the historical and literary background of the subject. His scientific papers were contributed mainly to the Institution of Electrical Engineers and to the *Philosophical Magazine*.

Thompson's studies of magnets and magnetic materials led naturally to the recording of hysteresis loops, dynamic as well as static, and to the harmonic analysis of the distorted wave forms associated with them, that is to say, the determination of the harmonic frequencies introduced in alternating magnetization by the non-linearity of the relation between magnetic induction and magnetizing force. This, presumably, gave the incentive to his development of a novel method of approximate harmonic analysis which he described in a paper to the *Philosophical Magazine* in 1911.

The particular virtue of his method lay in eliminating the intermediate operations of multiplying by sines or cosines and requiring only the averaging of selected ordinates.

It was, however, as an expositor of advances made by others that Silvanus Thompson was supreme. He excelled in the spoken word, the lecture demonstration and in the written description. In popular exposition it was his Friday Evening Discourses at the Royal Institution, and perhaps equally his Christmas lectures to children, which he allowed to tax his resources to the full and which exhibited his powers at their highest peak of performance. A notable example was the Christmas Lectures of 1896 given under the title *Light, Visible and Invisible*. These were published, in slightly expanded form, the following year. In these he pioneered the introduction of elementary wave theory into the teaching of geometrical optics. He also dealt with topics such as phosphorescence, fluorescence, cathode rays and the very recent epoch-making discovery of X-rays by Röntgen.

The range of his Friday Evening Discourses was wide, including, in 1890, *The Physical Foundations of Music*, in 1896, *Electric Shadows and Luminescence*, in 1902, *Magnetism in Transition*, in 1906, *The Electric Production of Nitrates from the Atmosphere* and in 1910 he gave a course entitled *Illumination Natural and Artificial*.

When Röntgen's discovery of the photographic action of X-rays was reported in *The Lancet* in January 1896, Thompson, who had himself been working in a rather similar way with cathode rays, reproduced the phenomenon immediately and he quickly became recognized as an authority on the subject. He discovered the photographic effect of a more deeply penetrating radiation produced spontaneously by uranium nitrate and reported this to Sir George Stokes. Stokes replied, inviting a communication for the Royal Society, but wrote again a month later saying:

> I fear you have already been anticipated. See Becquerel 'Comptes Rendus' for February 24th, p. 420, and some papers in two or three meetings preceding that.

In March, the Secretary of the Clinical Society of London wrote to Thompson, asking whether he could lecture upon and demonstrate Röntgen Rays to the Society and this he willingly did. A year later a number of medical men in London decided to form a Röntgen Society and they invited Thompson to become their President.

Thompson's interest in optics never flagged. although in middle life he could devote little time to the subject. 'The design of new polarizing prisms continued to be his favourite optical exercise.' The great distinction achieved in optical work in Germany excited his admiration and brought home to him increasingly the deficiencies in British optical practice and education. He became actively interested in the measurement of lenses and he devised a method for the determination of the optical constants of lens combinations of short focal length. This was communicated to the Royal Society in 1891 in a paper entitled *On the Focometry of Lenses and Lens Combinations*.

One of Thompson's aims was to improve the status and training of those engaged in the optical industry, in particular opticians. Some steps had already been taken by the organization of examinations for opticians by the British Optical Association and by the Spectacle Makers Company and by the granting of certificates by these bodies.

He determined to pay a visit to the famous optical works in Jena and in the spring of 1900 he went alone on holiday to the Thüringenwald. He was courteously received in Jena by Dr Abbe under whose direction the firm of Zeiss was operating. He met other leading men in scientific optics, Dr Schött, Dr von Rohr, Dr Rudolph and Dr Pulfrich, and he was shown testing and manufacturing processes in the glass works. In a letter dated 16 April, and written while visiting the Zeiss works, he noted with evident satisfaction:

> But now comes the most extraordinary thing. Abbe has turned the firm of Zeiss, together with his share of the Schott glass works, into a 'Carl Zeiss Stiftung', that is to say, into a sort of Socialist Company, in which all the work-people are co-operative owners. He who might be deriving an income of £6000–£8000 a year, simply draws a salary of £600 as do also his three co-directors. All the other profits are divided amongst the employees, and it is arranged that whenever Dr Schott dies, his concern shall entirely merge into the Carl Zeiss Stiftung.

Thompson returned to England fired with enthusiasm to improve the condition of the British optical industry.

Following the initiative of the British Optical Association, the ancient guild known as the Spectacle Makers' Company began to formulate a plan for a system of examinations. Plans were completed early in 1898 for a scheme of examination to be conducted by two members of the Company acting in association with a third 'external' examiner who should be a scientist of standing. The first to be approached, but without success, was Sir William Crookes. Next was Thompson, who accepted. These examinations in optics became a matter of deep personal interest and concern to him. Initially, the examinations were concerned mainly with the technicalities of lens design, manufacture and testing, no provision being made for sight testing or for the recognition of diseased conditions of the eye. When, after several years, the Guild decided, on the recommendation of Kelvin and other eminent men, to add a diploma in sight testing, Thompson regarded the development with distinct misgiving. He was seriously apprehensive of the dangers of abuse. However, in the end, the common sense of the matter prevailed. In practice, however much the ophthalmic surgeons might disapprove, unqualified sight testing had gone on from the earliest days and the situation was unlikely to be worsened by the availability of an educational qualification. Thompson retired from the post of chief examiner in 1908. He continued, however, to take part in the examinations and in 1911 he served, along with Richard Glazebrook, then Director of the National Physical Laboratory, as examiner for the Diploma in Applied Optics and Optical Instruments.

Walter Fincham, who in later life became the author of one of the most widely used textbooks on Ophthalmic Optics, recalls that he was examined by Glazebrook and Thompson. He was, as it happened, the only candidate for the Diploma. The written papers were followed by a searching oral examination. Fincham was first questioned by Thompson, and made, as he thought, a good showing. Then Glazebrook followed with a number of straightforward questions, but finally, to Fincham's dismay, which he could not conceal, asked how the candidate would measure the transmission of a pair of prism binoculars. To his astonishment he heard Thompson interject, 'Ah, Glazebrook, that's a most interesting question. I've often wondered how to do that. How do you do it in your laboratory?' Fincham passed!

Thompson was admitted to the Fellowship of the Spectacle Makers' Guild and he became in 1899 a Freeman of the City of London. He was also interested from its beginnings in the Optical Society, a body founded in 1900 for the discussion of scientific and technical aspects of optics. He became its president in 1905, and he chose for his presidential address the subject: *The Early Literature of Optics*, which he illustrated by volumes from his own library.

The second Optical Convention organized by the Society was held in London in 1912 at the Science Museum and Imperial College. Thompson was President, and in his address he gave a review of progress in the various areas of the field, emphasized the importance of associating theory with practice, and set the subject in historical perspective by describing the fundamental contributions of Newton and Huygens. In this connection he had undertaken, some time before the meeting, a translation of Huygens' *Traité de la Lumière*. This was published by Macmillans in time for the Convention and was sold to members at half the published price. Characteristically, Thompson was propagating the gospel of the value of history as a background to current technology.

Thompson's philosophy of technical optics was expounded in detail, under the title *Opto-technics* in an address which he gave to the Society of Arts in 1902. He asserted that there might be some 20,000 persons employed in the optical industry in London and he described the extreme paucity of appropriate educational provision. As early as 1886 he had drafted a scheme for a specialist department in the Finsbury Technical College and later he had hoped for the establishment of an Opto-technical Institute in Clerkenwell, the effective centre of the optical industry. He was highly critical of the lack of really good optical literature in English. Why, he questioned, did not the optical industry request that the money granted annually to the Technical Education Board of the London County Council but *not used* be devoted to the instruction of opticians? He was convinced of the need immediately to establish an Opto-technical Institute properly equipped and staffed and responsible for a 'respectable optical journal'. Despite sustained efforts, almost up to the time of his death, Thompson's Institute did not materialize, although substantial development did take place in specialized departments in Imperial College and the Northampton Polytechnic, now the City University.

It is almost certain that Thompson initiated the application of polarized light to the study of stresses in the members of engineering structures – a technique which achieved wide application under the title 'Photo Elasticity'.

Thompson recruited to his staff in 1904, as Professor of Mechanical Engineering, Ernest G. Coker, whose special field of interest was the determination of stress distributions in engineering structures. In view of Thompson's detailed and specialized knowledge of optical polarization it is highly probable that he suggested to Coker the possibility of using polarized light in demonstrating stress distribution in a transparent model of the member to be investigated.

Thompson and Coker produced two joint papers. The first entitled *Note on the application of polarized light to determine the condition of a body under stress* was a British Association Report presented to the Winnipeg Meeting of the BA in 1909. This paper begins, 'The application of polarized light to observe optically the strained condition of glass and other transparent materials when subjected to stress has been known since the days of Brewster and Biot.'

A second joint paper on *The Design and Construction of Large Polariscopes* was contributed to an Optical Convention held at Imperial College in 1912. Thompson did not pursue the matter further but Coker, in association with L. N. G. Filon, developed the subject exhaustively at University College. The technique achieved great importance and is only now being complemented and replaced by numerical computation.

Gas-lighting had been in use in public places and in private houses since the early 19th century and by Thompson's time had achieved wide distribution. Electric arc lighting had been increasingly utilized since the

11 *A lamp bracket designed by Thompson and installed in his home in Hampstead*

1870s and the incandescent electric lamp from the 1880s. By the turn of the century 'illuminating engineering' was beginning to be recognized as a branch of applied science and the Illuminating Engineering Society was established in 1909 with Silvanus Thompson as its first President. In the following year he delivered a Friday Evening Discourse to the Royal Institution on the subject of *Illumination, Natural and Artificial*. At his suggestion, the Society set up committees of enquiry into street, school and library lighting. An American Association of Illuminating Engineers had been established some three years earlier than its British counterpart. The initiative was taken by the American Association to hold an International Conference on questions of Photometric Nomenclature and Standards and the conference was duly held in 1912. This fostered the development of the movement to establish international standards for illumination associated with the work of the International Commission on Illumination.

Scientific Internationalist

Silvanus Thompson owed his academic qualifications to the facilities provided by the University of London. His interests in education, which were in a very real sense catholic, had developed more in the technological than the academic sphere, but his loyalty to his university and his concern for its proper development were deep and sincere. From its foundation in 1836 until 1858 the graduates had no voice in the councils of the university. 'An autocratic Senate appointed by the Government was in supreme control'. Ten years of controversy between a Graduates Committee and the entrenched Senate resulted in May 1858 in a new Charter embodying provisions for Convocation – the graduate body.

By the time Thompson took up his appointment at Finsbury, a movement had developed in Convocation to establish teaching, as distinct from examining, as a major function of the university. Other schemes had already been promulgated for the 'Promotion of a Teaching University for London'. Thompson immediately associated himself with the Convocation project. One feature of the project was, however, unacceptable to Thompson, and that was the strengthening of the Crown-nominated Senate at the expense of the graduate body. An alternative scheme, more in line with Thompson's thinking, was accepted by Convocation but was later rejected by the Senate.

The twin questions of the possible reorganization of the university as a teaching institution and the alternative of establishing a second and teaching university, leaving the University of London as an examining body, became a major issue. Two Royal Commissions considered the problem between 1892 and 1898. *The Times* published several long articles and leaders on the subject and much correspondence to which Thompson contributed. He read, to the Royal Society of Arts in January 1896, a paper entitled *The Making of a Great University*. In this he stated what he considered the basic questions. 'It is therefore from the point of view of the scholar, whether he be nominally student or nominally teacher, that the claims of a university to be considered great must be decided. Does it bring him into an atmosphere of mental activity and progress? Does it afford contact with living thought? Does it give the stimulus of intellectual struggle so essential for improvement of knowledge? Does it furnish the means and appliances of learning? Does it provide the scholar with libraries, and give him access to the mental furniture of the past and of the present? Does it offer to the investigator the means of pursuing research? If it does, then it is fulfilling its functions as a University. The test of greatness is to be found in the degree to which it thus ministers to the intellectual progress of the age.'

In August 1898 the Act, representing the work of the second Royal Commission, received the Royal Assent and established, in large measure, the principles for which Thompson had struggled so hard. In October 1900, Thompson was elected to the Senate as a Convocation member.

Thompson had clear and definite opinions as to the separate and distinctive roles of Technical College and University. In *A Letter to a Graduate*, written at the height of the university reorganization controversy, he commented with reference to Finsbury, 'No one has ever proposed that the Technical College should become a constituent college. If anyone were to propose it, I, as its educational head, should oppose the suggestion to the uttermost. The Technical College is a technical, not an academic, institution . . . the training it gives is professional rather than scholastic. To make it a University College would be entirely to change the character of its training and divert it from its present useful though less ambitious work'.

Despite his deep interest and involvement in technical college work he would, in 1901, still have been willing to transfer to an administrative post in the University of London. The newly created post of Principal Officer was to be filled. He wrote to the Vice-Chancellor, Sir Henry Roscoe, indicating his willingness to be considered. From a short list of four, which included Thompson, Professor Arthur Rücker was appointed. It is probable, since the appointment was a Government one, that Thompson's outspoken criticisms of British Imperialism and the conduct of the Boer War militated severely against his being selected.

On the Senate, Thompson found himself out of sympathy with the prevailing imperialist attitude and was often a member of a small minority. He was not re-elected at the end of his term. Notwithstanding this, he allowed his name to go forward again in 1907 but was unsuccessful. Thompson felt very deeply frustration and disappointment at his inability to influence the policies of his university. In his Society of Arts paper he had written: 'Prejudice, ecclesiastical bias and ignorance all rear opposing heads. Already the ignoble methods of party politics, seeking to catch the support of the least informed, have been resorted to, to prevent the calm and dispassionate discussion of the questions. Common fairness has been laid aside, not once nor twice, in the attempt to stop the wheel of progress.' These influences, persisting after the reorganization, disheartened Thompson almost to the point of embitterment. For the rest of his life he took little active part in the affairs of the university, other than serving on the Library Committee. Following a proposal he made on the committee in 1907, it was decided to create a strong bibliographical section, and this has become a most important and valuable feature of the library, serving as a basis for information services.

In 1889 the Society of Telegraph Engineers and Electricians changed its title to that of the Institution of Electrical Engineers. Thompson had been elected to membership of the Society in 1882. He read his first paper to the Society, entitled *Remarks on Contact Resistance*, in 1883 but he could not, until he moved to London in 1885, attend meetings with any regularity. Thereafter he appeared often and he became one of the most lively and cogent contributors to discussion. He was elected to the

Council of the Institution in 1886 and in 1899 to the Presidential Chair.

In electrical science and engineering it was an exciting time. Hertz had demonstrated electromagnetic waves in 1886–7. Sir William Thomson and Oliver Heaviside were contributing notably to the theory of electrical transmission. Ferraris and Tesla were producing two- and three-phase alternating currents. Ferranti was transmitting alternating current power by cable at 10,000 volts and by 1891 Parsons was building turbo-alternator sets of 100kW.

The generation and distribution of direct current was well established by 1890, but already alternating current systems were beginning to demonstrate their superiority. What came to be known as 'The Battle of the Systems' had begun. Speaking in a discussion at the Institution in 1891, Sir William Thomson* confessed to a belief that, 'the direct current system is destined to predominate and ultimately to supersede alternating currents for all densely populated districts'.

In 1894 the British Association met in Oxford and Silvanus Thompson presented to Section G a paper entitled *Some Advantages of Alternate Currents*, which has been described as 'the most complete and masterly vindication of the alternating current system'. He concludes 'with such possibilities open in the future for alternate current working, and with such advantages in respect of motor power over continuous current working, it can hardly be doubted that, save in a few special cases, the vast majority of central stations will henceforth be operated by alternating current.'

Thompson's international contacts were, all the time, being maintained and extended. In 1891 he became a member of the Societé International des Electriciens and was elected Vice-President of the Electrical Exhibition at Frankfurt which was to be the venue of the Congress in that year. It was here that a pioneer demonstration of alternating current transmission was given. Three-phase power was successfully transmitted to Frankfurt at 25kV from the hydro-electric station on the Rhine 110 miles away. Thompson read, in German, a paper on alternating currents.

In May 1899 Thompson became President of the Institution of Electrical Engineers. Undoubtedly the continuing hostilities of the Boer War distressed Thompson and his family and diminished in some degree their enjoyment of the felicities of the office. Typically, one of his first actions was to make a proposal to promote, even if in only a small way, international understanding. He proposed that the Institution should make a summer visit to Switzerland. The visit started in Basle, took in several engineering works, the power station at Rheinfelden, and the construction work on the Jungfrau Electric Railway, which at that time had been completed as far as the Eiger Glacier. From Switzerland the Thompsons went on to Italy, where Silvanus was to represent the Institution at the

---

* Created Lord Kelvin of Largs in 1892.

celebrations in Como of the centenary of Volta's great discovery of the principle of the Electric Pile. Earlier in the summer an exhibition of Volta's apparatus, manuscripts and relics had been opened in Como but, tragically, the building had been destroyed by fire in July and much important material lost. Fortunately the original pile was saved. Following the fire, Thompson had written at once offering the loan of manuscripts and letters from his own collection. The re-opening of the exhibition coincided with a meeting of the International Electrical Congress. To this Thompson contributed a paper, given in Italian, on Magnetic Images.

Thompson's Presidential Address was given in November 1899. In this he gave a comprehensive and critical review of the state of the *art*. He commented on the unfavourable position of inventors, suggesting that an inventor requiring, say, £10,000 or £25,000, to put his invention on the market 'must either raise the necessary capital by personal effort amongst his own friends or he must put it in the hands of a promoter who will want to bring it out with a capital of at least £100,000. In other words, it is almost impossible to find a middle course between a purely private affair and an over capitalised swindle'. He referred to the foundation of the National Physical Laboratory and to the publication by the Institution, jointly with the Physical Society, of Science Abstracts. He commented on the future development of the Institution, mentioning the recently instituted policy of establishing local branches and overseas centres. Ironically, in the light of subsequent events, he made a gracious acknowledgement of the generous gift by Henry Wilde, FRS, of £1500 to establish a special benevolent fund. The gift had been made in recognition of Wilde's election to Honorary Membership of the Institution.

He referred also to the unique event of the reading of a paper by a woman scientist, Mrs Hertha Ayrton, on the subject of the hissing of the electric arc, and expressed pleasure at her subsequent election to membership as the first woman member of the Institution.

As has been mentioned, Thompson became a member of the Physical Society in 1875 and read to the Society, in the following year, his first scientific paper. Over the years he made contributions from time to time which were published in the *Proceedings* or in the *Philosophical Magazine*.

An important step was taken by the Physical Society and the Institution of Electrical Engineers in 1897 in their joint sponsorship of the publication of *Abstracts of Physical Papers* which, later, became *Science Abstracts*. In this Thompson was involved as a member of a joint Committee of Management for the new publication. His colleagues were W. E. Ayrton, James Swinburne, R. E. Crompton and J. S. Raworth.

Thompson was elected to the Presidential Chair of the Physical Society Society for the session 1901-2.

Thompson was elected President of Section (G), Engineering, of the

British Association and delivered his presidential address at the Leicester meeting in 1907. Reflecting activity and special interest in various areas of the engineering field, there were, among the subjects discussed in the Section, gas and petrol engines represented by papers by Dugald Clark and Bertram Hopkinson, tuning in wireless telegraphy by Oliver Lodge and Pupin's compensated cable for telephone transmission by Preece. Thompson's address first reviewed the way in which electrical science had been applied, almost universally, to meet human needs. He went on to develop the thesis that, in turn, technical development can facilitate and even initiate further advances in pure science – what, in fact, Ewing called 'the beneficent reaction of practice on science'. Finally, he treated at some length the subject of engineering education and training, commending highly the report on these matters issued the previous year by the Institution of Civil Engineers.

One of the most interesting papers read by Thompson to the Institution of Electrical Engineers was prepared by him at the request of the Papers Committee. The subject *The Aims and Work of the International Electrotechnical Commission* was one upon which he was almost uniquely qualified to speak in virtue both of his service on the Commission itself and of his wide knowledge of science and technology in overseas countries. In the paper it is stated that 'The impulse towards the unification of electrical measures came first through the practical submarine telegraph pioneers, whose methods of electrical measurement were, even in the fifties, far more scientific than those of the professed men of science.' The International Congress of Electricity held in Paris in 1881 was the first of its kind and at that meeting the name of the unit of current was changed from Weber to Ampère. Differences in terminology grew up with manufacturing industry and disputes about specifications and contracts tended to arise from such differences and from want of precision in definitions.

The *Physicalische Techniche Reichsanstalt*, the *Laboratoire Central d'Electricité*, the Bureau of Standards and the National Physical Laboratory had all been set up by the turn of the century and in 1901 the British Engineering Standards Committee was formed.

It was at the St Louis International Congress of 1904 that the Chamber of Delegates resolved 'that steps should be taken to secure the co-operation of technical societies of the world by the appointment of a representative commission to consider the question of the standardisation of the nomenclature and ratings of electrical apparatus and machinery'.

Referring finally to the establishment of Local Electrotechnical Committees and to the work on nomenclature done by them, the paper, not unexpectedly, finishes with the words 'an important factor in furthering the peace of the world'.

At the end of April 1908, Thompson delivered at the Institution of Electrical Engineers the first Kelvin Lecture established by the Institution in commemoration of the life and work of the great man, who had died the previous year.

Matters Legal – Scientific

From the time of his professorship in Bristol, Thompson found his services in demand as a consultant and an expert witness. In his earlier days the remuneration was most useful, as a supplement to salary. His first experience as an expert witness was in an action concerned with telephone apparatus. Thompson had himself filed a patent application in 1882 for an improvement in telephone apparatus, and his involvement in a consultative capacity stimulated him to apply his inventive faculty further in this connection. The master patents covering the electromagnetic receiver and the carbon-granule transmitter were held by the Bell-Edison combination, operating in England under licence from the Post Office, as the United Telephone Company. There was every incentive to invent a device, transmitter or receiver, which did not infringe the Bell-Edison patents. In 1884 Thompson patented a form of transmitter which, it was believed, would avoid such infringement. The essential point of the invention was that it did not employ a diaphragm. A variable resistance carbon contact was created at the contact edges of a form of flap or ball valve. Although not a particularly elegant device, it must have worked well for it was quickly taken up and a company formed to exploit it. No sooner were several of the instruments in service than the United Telephone Company brought a successful action for infringement. The judgment was upheld on appeal, Lord Justice North making the remarkable statement that 'every surface which can vibrate is a diaphragm'. The company, of which Thompson was a director, was forced into bankruptcy.

This setback did not discourage him from further invention in the telephone field. In 1892 he obtained a patent on a means of equalizing the transmission characteristics of telephone cables by inductive *shunt* loading. The idea, although sound in principle, was never exploited, for the improvement in articulation was inevitably associated with additional attenuation, much greater than that produced by the method of series inductive loading devised by Pupin very shortly afterwards.

In 1893 Thompson served as an expert witness for the defendants in an important patent action which came to be known as the 'Three Wire Case'. John Hopkinson, Consulting Engineer and Professor of Electrical Engineering in King's College, London, brought an action for alleged infringement of his 'three wire' patent against the St James & Pall Mall Electric Light Company. The action was successful. In his summing up, in the Chancery Division of the High Court, Mr Justice Romer made an interesting reference to Thompson's evidence. It was asserted by the defendants that the principle enunciated in the patent was previously known, although it had not been applied in that particular form, and in consequence was of no real novelty. In cross-examination it transpired that Thompson had, shortly before, given a lecture dealing with systems of electricity supply and had *not* included, in the systems he described, the three wire principle!

Thompson had the misfortune in 1902 to be defendant in a libel action which originated during his preparation of the sixth edition of *Dynamo Electric Machinery*. One of the most notable contributors to the development of electrical generators in the 1850s–1860s was Henry Wilde, FRS. During this period it was demonstrated that, for the purpose of generating, mechanically, an electric current, the field of an electromagnet could be employed in place of that of the permanent magnet which had been used up to that time. Thereafter, for power generation, the dynamo-electric machine replaced the magneto-electric generator. This great step was made independently and nearly simultaneously by several inventors, Henry Wilde, Werner von Siemens, C. F. Varley and Charles Wheatstone and Farmer. Wilde was the pioneer. Using a small magneto generator to produce the current to excite an electromagnet which constituted the field magnet of a larger machine, Wilde was able to show that, by repeating the process, an 'indefinably small amount of magnetism is capable of inducing an indefinitely large amount of magnetism, or further, of dynamic electricity'. Wilde patented the method of using a magneto generator to excite a dynamo in 1863. Self excitation from the residual magnetism and the shunt and series field connection were the subject of almost simultaneous invention nearly four years later.

Wilde's contribution was, in due course, generously acknowledged. He was elected to Honorary Membership of the Institution of Electrical Engineers in 1901 and he had been awarded the Albert Gold Medal of the Royal Society of Arts in the previous year. Wilde was, however, dissatisfied with the terms in which the citations were made. They did not describe him as the inventor of the dynamo which he considered himself to be. It was largely a matter of terminology but one in which Wilde became obsessed by a sense of injustice. Silvanus Thompson became the victim of Wilde's displeasure by virtue of his being the author of the authoritative work on dynamo electric machinery which was about to appear in its sixth edition. Following an exchange of letters with Wilde, Thompson did agree to some modification of his reference to Wilde's work but his integrity and his meticulous care in the matter of verbal accuracy would not permit him to credit Wilde with 'the invention of the dynamo'. Wilde took the extraordinary step of initiating an action for libel against Thompson. Thompson's distress was acute but he could not, in honesty, compromise further in the matter. The case was argued before Mr Justice Buckley on 12 March 1903. Dismissing the action the judge commented, 'It would be an evil day if it were the law, that, if one man made a concession to another man for the sake of peace, the result is that there is a contract upon which the one can sue the other for specific performances or damages.' The case was carried by Wilde to the Court of Appeal where, again, the action was dismissed.

For Thompson, perhaps the happiest outcome of any legal action in which he was involved was that of Lodge's application in 1911 for the prolongation of his patent No. 11575 of 10 May 1897, entitled *Improvements in Syntonized Telegraphy without Line Wires*.

By 1911 wireless telegraphy had been established throughout the world, chiefly as a result of the resource, ingenuity and determination of Guglielmo Marconi. Apart from the major advance – the development of the aerial-earth system – Marconi did not contribute to the scientific principles of the subject. One of the most important concepts was that introduced by Lodge as 'syntony', that is, tuning. Lodge had demonstrated tuning in 1894 and had enunciated the basic proposition that a closed resonant circuit will, after shock excitation, sustain its oscillations for a relatively long period but will radiate little, while the oscillations of an 'open' circuit will radiate strongly but die away quickly. The essence of Lodge's patent was the introduction of additional variable inductance

12 *A Branley coherer. This was given to Thompson by Oliver Lodge. The coherer was invented by Branley, French pioneer of wireless telegraphy, and is a detector of radio telegraph signals*

into the oscillatory circuit thereby lengthening the wave train sufficiently to enable tuning to be carried out. This then introduced the frequency selective feature into transmission and reception.

The extension of this master patent, was, naturally, strenuously opposed by the Marconi Company and other interests. Thompson's close and long-standing friendship with Lodge and his deep sense of justice in matters of attribution, virtually necessitated his support as an expert witness. The notes which he prepared before the action were subsequently published and in a brief foreword Silvanus comments:

> It is a matter for congratulation that the clear and able judgment of Mr Justice Parker, in granting the prolongation for seven years, had so thoroughly established the claims of Sir Oliver Lodge as the pioneer in tuned radio telegraphy.

Biographer and Friend

Thompson's interest in magnetism was almost life long. His first published paper on the subject dealt with magnetic figures, that is to say, iron filings pictures. His last, published a year before his death, was on *The Criterion of Steel suitable for permanent magnets*. At a very early stage he had been fascinated by the historical background of the lodestone, the magnetic needle and finally electromagnetism. The index of the first edition, 1881, of *Elementary Lessons in Electricity and Magnetism* contains no less than ten references to William Gilbert, of Colchester, whose famous treatise *De Magnete* was published in the year 1600. Thompson gave a number of popular lectures on magnetism, two of which were Friday Evening Discourses to the Royal Institution. His own researches provided useful data on the self demagnetizing coefficients of bar magnets and he applied another of his research interests, harmonic analysis, to the determination of the harmonic components of hysteresis loops. Probably his greatest contributions were, however, made in the literary and historical fields.

In 1882 Thompson acquired a copy of Gilbert's great work *De Magnete Corporibus, et de Magno Magnete Tellure*, published by Peter Short in London in 1600. Gilbert and his work were to provide him with the stimulus and the opportunity to exercise all his scholarly skills in their highest perfection. Elected in 1890 to membership of the 'Sette of Odd Volumes' he contributed his first Opusculum, or little work, the following year under the title *Gilbert of Colchester, An Elizabethan Magnetiser*. In the same year he wrote and had privately printed a small volume dealing with the life and work of William Sturgeon, the remarkable, self-educated 'electrician' who constructed the first powerful electro-magnet.

One of Thompson's friends, Conrad W. Cooke, had in 1889 written an article in *Engineering*, pointing out that no translation of Gilbert's *De Magnete* into any modern language had been made, that the tercentenary of the publication of the work would occur at the turn of the century, and added that a Gilbert Club had been formed for the specific purpose of translating and producing the work in as nearly as possible its original format, in celebration of the event. Sir William Thomson (later Lord Kelvin), accepted the office of President of the Club and prominent members of the Royal College of Surgeons, the Physical Society, the Royal Institution and the Institution of Electrical Engineers offered support. The work of translating various parts of the book was undertaken by ten different persons, two of whom, Thompson and the Reverend W. C. Howell, performed the task of revision and correction for the press. *De Magnete* was beautifully printed and produced by the Chiswick Press. Two hundred and fifty copies were printed for distribution to subscribers.

For a man of Thompson's width of interest, scholarly curiosity and didactic impulse the work of translating and preparing for press a finely printed folio edition of *De Magnete* must have been enormously rewarding. Philology, typography, and the study of papers for printing, all fascinating subjects in themselves, could hardly match the excitement of finding the first authentic signature of Gilbert in the Public Record Office, then identifying others in the library of St John's College, Cambridge, where Gilbert had studied and, later, through the good offices of his bookbinder friend, Douglas Cockerell, being able to buy Gilbert's own copy of Aristotle. Thompson also studied the pedigree and arms of the Gilbert family, searched for a genuine portrait and identified the device upon the title-page of *De Magnete*, of a serpent entwined around a T-shaped support, as the mark of the printer, Peter Short. Then he wrote explanatory *Notes on the de Magnete of Gilbert to accompany the Gilbert Club translation*. At a later date he produced, privately printed, a small book entitled *Gilbert, Physician*, dealing with the doctor's professional life.

A further product of his work on Gilbert was a translation of the very early writings on magnetism of Petrus Peregrinus, a soldier of fortune of the thirteenth century, whose manuscript was entitled *Epistola de Magnete* and dated 1269. The publication of the Gilbert translation was, in fact, anticipated in America. Dr P. Fleury Mottelay began, also in 1889, to work on a translation and this was published in New York in 1893.

In 1913 an International Congress of Historical Studies was held in London. Thompson represented the Royal Society and contributed a paper on 'the Origin and Development of the Compass Card'. The early compass card was known as a wind rose, or *Rosa Ventorum*. Thompson's paper was published in the Proceedings of the British Academy, under the title 'The Rose of the Winds'.

Thompson's first venture in biography was his life of Phillip Reis, which was principally a defence of Reis's claim to be the first to produce a practical telephone. Of quite a different character and quality were his *Life of Faraday* and his *Life of Lord Kelvin*. The Faraday biography was

*En suit ladicte figure pour cognoistre ce que dessus, Aussi
les noms & tyns des Vents.*

13   *A compass card from S. P. Thompson's book* The Rose of the Winds
      (see page 48)

undertaken at the invitation of Sir Henry Roscoe, who at that time was editing a series of short biographies of scientists for Cassell. While Thompson's admiration for Faraday was unbounded, he was nonetheless objectively critical. He had an unrivalled first-hand knowledge of Faraday's experimental researches, since he had performed at the Royal Institution many of the experiments, using Faraday's own apparatus. The result was an assessment at once accurate and delicately sensitive.

The biography of Kelvin was a major work of scholarship. It is even today still the standard work. Thompson was introduced to Kelvin (then Sir William Thomson) at an Inventions Exhibition in Glasgow in the spring of 1876. Later the same year he met Thomson again, at the British Association meeting, also held in Glasgow. Fifty-two years of age, recognized as one of the world's leading scientists, with numerous distinguished contributions to his credit in physical science and its engineering applications, Thomson was remarkable for the encouragement he gave to young men at the outset of scientific careers. Silvanus Thompson personally experienced this encouragement, and he developed both admiration and affection for the great man. The younger man's sincerity, ability and enthusiasm must, on their part, have touched a sympathetic chord in the older man, for, as time went by, a real friendship was formed.

Meetings between Kelvin and Thompson became more frequent as the latter achieved greater seniority and influence in the various scientific and engineering societies. In Thompson's year of office as President of the Institution of Electrical Engineers (1899) he had the pleasure of inviting Kelvin to become the first Honorary Member of the Institution. The satisfaction Thompson had obtained from writing his biography of Faraday, and the high praise the book had received inspired him to approach Kelvin and suggest that he should undertake his biography. He waited, quite probably with some trepidation, for Kelvin's reply to his letter. When it came, he was delighted to read: 'As you kindly told me you had been thinking it possible you might wish to undertake writing an account of my own scientific work, I can say I would feel complete confidence that, in your hands, it would experience thoroughly satisfactory treatment. If you are inclined to talk over the matter just now, shall we meet one of these days, at any time that would suit you – either morning or about tea time after the working day is over?'

Thompson's approach had been made none too soon. Eighty-two years of age, mentally active and with a memory little, if anything, impaired, Kelvin was nonetheless in poor health and he had little more than a year to live. Quite a number of 'sittings' took place in the spring and autumn of 1906 and several in the spring and early summer of 1907. In the autumn Kelvin died.

Thompson's plan for the book had to be drastically altered. Instead of being one volume, containing only a few letters and treating scientific matters in a way suited to the non-mathematical reader, it now became a full-scale treatment of Kelvin's scientific work, together with a well-documented account of his life in its social setting. Almost everything except Thompson's responsibilities at Finsbury and other inescapable engagements, was set aside to tackle the task.

Thompson was engaged in almost endless correspondence and consultation in obtaining and checking information on scientific and personal matters. The chapter on thermodynamics which he found particularly difficult, he sent to Lodge for criticism. He was fortunate in making contact with J. D. Hamilton Dickson, a Fellow of Kelvin's old college, Peterhouse, Cambridge. Dickson had been one of Kelvin's students and an ardent admirer. From him there came a veritable flood of personal reminiscence and of information from College sources. The biography was published by Macmillans in 1910 in two volumes. The book was very favourably reviewed, was accepted as 'the standard authority' for Kelvin's life and has so remained.

Macmillan also published in 1910 a small volume entitled *Calculus Made Easy*, the author of which had assumed the pseudonym 'F.R.S.' The secret was kept until after his death. The author was Thompson.

Over a long period Thompson had exercised all his skill and ingenuity in teaching his students the elements of the differential and integral calculus. A branch of mathematics essential for the study of engineering, the calculus was at that time considered a difficult subject, a *pons asinorum* crossing the divide between elementary and advanced studies. Poking fun at orthodox mathematicians who made the introduction to the calculus excessively difficult, the book presents the elements of the subject in the most lighthearted and informal way. The concluding paragraph reads:

> 'There are amongst young engineers a number on whose ears the adage that what one fool can do another can, may fall with a familiar sound. They are earnestly requested not to give the author away nor to tell the mathematicians what a fool he really is.'

These words must have been his rallying call to his students at Finsbury! *Calculus Made Easy* has carried Silvanus Thompson's name throughout the world wherever engineering science has been and is taught. The book has been reprinted many times, most recently in 'updated' form as a paperback, and has served as an inspiration and kindly guide to countless young engineers and scientists.

Thompson's writing exhibits not only his power of lucid and accurate exposition and felicity of expression but also his meticulous care to use words conformably with their etymology and traditional usage.

In the discussion following his paper to the Institution of Electrical Engineers on Rotatory Converters, he remarked: 'Lastly let me protest against the insinuation that in using the good old adjective "rotatory" I have altered the English Language. The old English language has many adjectives like "rotatory, explanatory, inflammatory, sanatory" and "undulatory", but I do not think that those adjectives would be improved by cutting out what might seem an unnecessary syllable.' He was consulted from time to time by Dr James Murray, editor of the Oxford English Dictionary, which was then in preparation and he, in turn, would refer difficult questions to Murray as the leading authority.

From his boyhood activities in sketching and caricature, Thompson had slowly and assiduously cultivated the art of water colour painting. After his visit in 1883 to Germany to collect information for his *Life of Phillip Reis*, he went with his wife to Switzerland. They greatly enjoyed three weeks in the Oberland where for the first time he made successful sketches of glaciers and snow peaks. On later Continental holidays he exercised his increasing skill. By the mid-1890s he was a member of the Royal Water Colour Society Art Club and was contributing to the Society's annual autumn exhibition. Probably in the exercise of his skill in painting

14 *Walderswick Village, by Thompson*

he enjoyed greater serenity of mind than in any other activity. He was particularly successful in painting ice and in his treatment of glaciers.

Thompson was born into the Society of Friends, as both his parents were Quakers, and he remained so until the end of his life. This did not mean however that he accepted without question all the tenets and practices, some narrow and illiberal, which were common in Quaker circles in his younger days. In 1895 he contributed to a conference of the Society of Friends held in Manchester a paper entitled *Can a scientific Man be a sincere Friend?* In this he asserted forthrightly that there could not be any conflict between science and religion. 'That which is divine truth,' he said, 'modern thought will leave wholly untouched or will touch but to confirm'. It has been said that by his sustained development of the argument he succeeded, in large measure, in achieving the acceptance of this principle in modern Quakerism. During the last ten years of his life he worked, as opportunity offered, on a book entitled *A Not Impossible Religion* which he left unfinished and which was published posthumously in 1920. He ends his introductory statement with the words, 'The author publishes this work with the conviction that no advance in religious thought is possible unless the quest for truth, without fear of the consequences to accepted tradition, be ever accompanied by at least an equal regard for the preservation of a reverential spirit.'

It is interesting to speculate whether, with his superb intellect, Thompson could have made some really notable contribution to science had he been willing to limit his range of interests.

Perhaps his most outstanding personal characteristic was his genius for friendship. His capacity for work was prodigious and even by the conservative measure of his output of publications must represent a life regulated as to the utilization of time. Yet he found time for his family, for his social contacts, and, most astonishing of all, for an immense circle of friends. His friendships were catholic in the true sense of the term and were unimpaired by the sharp rebuke or caustic comment which he would make, if he thought the occasion demanded, usually in writing rather than in speech. Even the prickly Oliver Heaviside wrote, in the most friendly but characteristic way, about the Kelvin biography:

> Dear Thompson, I received your magnificent and expensive work yesterday and at once set to work to read it. Opening the book at random, the very first thing I saw was a printer's error, n for u or the other way. After this bad beginning, I proceeded to read page after page. Astonishing incoherency? Careful examination revealed the fact that the pages were uncut. Spent the rest of the day cutting it. Finished 11.30 p.m.
>
> Started again this morning. What an unbalanced work it is. One volume is much bigger than the other! Thought of Sterne's critic. Read a lot here and there in Vol. 2. Found I was considered to be a Nihilist, never heard of that before. Not altogether wrong though. I have thrown bombs occasionally, and they exploded, too, and did some damage to the grand old man's views. Found out why he did

15 *A glacier in the Alps, 1898, by Thompson*

not like 'curl'. He broke his leg while 'curling'! Who can wonder? The vector, too, never any use to anybody! What a story! Found letter to me. Did he keep a draft, I wonder.

He was a fine old boy. When I converted him to my doctrine about self induction, he wrote me a nice letter expressing his appreciation of the very practical way I had reduced my theory to practice. In my reply I directed his attention to a paper in the Phil. Mag. just coming out on the conversion of the static field of a charge continuously to a plane wave by m^n of the charge. But I don't think he took it in for in later work he seemed not to have seen its significance. He never really understood the generation of electromagnetic radiation until my Nature letter appeared.

There are a lot of things in Vol II period 1880 to death, which need a little correction from the point of view of my garrets in the skies, imperfectly historical and popular errors etc. But it will take me a long time to go all through the book for which I am thankful.

Yours truly,
Oliver Heaviside.

There is an additional marginal note referring to Fitzgerald and a postscript longer than the letter itself referring to financial problems relating to his home, followed by a paragraph discussing the merits of a polish which 'I shall try a little on my old bike . . .' and commending Brunswick black as a boot polish.

It was during the Boer War that Thompson was elected to the office of President of the Institution of Electrical Engineers. To his astonishment – and acute embarrassment – he received from Major (later Colonel) R. E. B. Crompton a letter suggesting that, for the Annual Conversazione of the Institution, a Guard of Honour be mounted by the Volunteer Corps of Electrical Engineers. He replied: 'I am quite sure that your letter of the 20th instant was intended to be for the good of the Institution and not for the purpose of compromising me. . . . I hope you will not put me in the painful position of having to oppose any suggestion that might emanate from you'.

Thompson's friendship with Lodge had begun in early manhood and throughout Thompson's life they discussed matters scientific, social and religious with complete freedom and mutual trust. They would send to one another for criticism, drafts of articles prepared for publication and copies of their published papers and books for comment. During the First World War, Lodge sent to Thompson the draft of an article which was published in *The Daily Telegraph* under the title *Science and Industry*. In it Lodge makes some sweeping criticisms of British Patent Law. On these Thompson comments favourably and adds a number of observations, of which the following is typical:

'On one point that you don't touch you might at least strike a spark. I refer to the abuse of patents to deceive. The clever rascals of patent lawyers draft the specifications so that they can't be worked without bringing in expert knowledge that is carefully concealed in the text ... Also a lot of the patents for synthetic drugs Novocaine, Papinyme, Aspirin – are so drawn that they do not reveal how to make these things: if you follow precisely the instructions of the specification you fail to make the stuff, the specifications being fraudulent.'

In August 1915, Methuens published a booklet by Lodge entitled *The War and After*. On receiving a copy Thompson wrote a long letter, first recording a few misprints and going on to detailed criticisms of which a few may be quoted:

p. 34   last line. Who is 'our more brilliant Nietzschian prophet? surely not G B S. – the rank poseur.

p. 43   The first sentence is an unwarrantable assumption; exactly on a par with the self justification of the Inquisition. It is a most dangerous doctrine and is held – on their side – by the Prussians.

p. 120   'Deeds are the test of faith, actions the test of doctrine'. This is a sonorous aphorism, but what is the sense of it, *doctrine* is not *faith*, but what *actions* are not *deeds*?

p. 207   'Prussianism must cease'. Yes, but so must some other isms. Blatchfordism, Beresfordism, Churchillism, Kitchenerism.

Now I go back to Chapters XV and XVI. I don't think you make a sufficient distinction between non-resistance of personal injury and non-resistance of the community. I mean this: It is right for the community to step in and resist the doing of evil to the individual. It is wrong for the individual to take the law into his own hands and become judge and executioner where he is an interested party. There is the essential difference between lawful force and lawless force. All war is lawless. . . .

The man who could lay waste every farm from the Vaal to the Orange River seems to me as great a monster as he who laid waste the country from the Humber to the Tyne and while we rightly execrate the memory of William of Normandy we cannot call our hands clean when we commit our destinies to Kitchener of Khartoum. This is perhaps not the time to say so openly, but it is burned in upon my soul. *Our hands are not clean.*

Silvanus Thompson lived his religion. Perhaps the greatest tribute to the quality of his life was to be found in the memories of his Finsbury students who, in old age, would express admiration for his teaching, acknowledge his influence and speak with emotion of his kindness. One was Howgrave-Graham who recalled a Saturday in June 1916, when Thompson was in College after making representations unsuccessfully on behalf of one of his assistants appealing for exemption from military service. He was flushed and upset. On the following Monday he died.

Few men have been able to comprehend, within a lifetime, a greater range of achievement and contribution than Silvanus Philips Thompson, scholar, teacher, artist and Friend.

16 *Silvanus Thompson at his desk*